FLORA OF TROPICAL EAST AFRICA

ISOETACEAE

BERNARD VERDCOURT

Truly aquatic plants or terrestrial in seasonally boggy habitats. Stem short, corm-like, 2–4-lobed or unlobed, or better developed, elongate or rhizome-like; roots numerous, branched. Leaves/sporophylls* in a rosette, terete or subterete becoming flattened or spathulate at the base, with membranous margins and a delicate triangular ligule on the adaxial surface at the point where leaves narrow above the sporangial cavity. Sporangia usually of two sorts (although microspores and megaspores occasionally found in one sporangium) the microsporangia usually restricted to innermost sporophylls; sporangia sunk into the cavity, the wall thin, sometimes covered with tissue (velum) which can be entire, partial or reduced to a narrow or wider rim around the leaf at the margin of the cavity; where the velum is complete or nearly so there are therefore two layers of tissue over the spores. Megaspores trilete, often of two sizes (sometimes in other parts of world even more polymorphic), tetrahedral with conspicuous triradial and equatorial ridges, the faces variously verrucate or reticulate (or with various other sculptures in other regions); microspores monolete, bilaterally symmetrical, frequently with a conspicuous apical ridge. Gametophytes produced within the spore and often within the sporangium; sporophyte stage often produced by apogamy without a male gamete.

A single very widely distributed genus is now recognised with about 130 species [see Jermy in Kubitzki, Fam. Gen. Vasc. Pl. 1: 26–31 (1990)]. *Stylites* Amstutz restricted to South America and formerly recognised as a second genus of the family is now considered to be a subgenus – subg. *Stylites* (Amstutz) L.D.Gómez.

ISOETES

L., Sp. Pl.: 1100 (1753) & Gen. Pl. ed. 5: 486 (1754)

Description as for family.

The species are difficult but may become easier when modern techniques have been used. Collectors need to describe the lobing of the corms, variation of the dimensions of the plants in a population, colour of the sporangia, shape of the basal scales and to make sure mature spores are present. Collections in spirit are needed so that detailed morphology and histology can be carried out which may help to clarify the chaotic taxonomy.

I have recognised only 6 species in East Africa.

Various terms are used in the literature to indicate the surface areas of the megaspores. The large hemispherical surface is referred to as the apical, upper or distal face and the 3–(4) small surfaces form the dorsal, lower, basal or proximal face. I have avoided this confusion by referring to the large face and the 3 smaller faces which is obvious by inspection.

---

* A sporophyll is simply a leaf bearing a sporangium, and throughout this account the term leaf is used for both.

1

1. Velum complete (i.e. there are two coverings between the adaxial face and the spores, a very thin sporangial wall and a thick layer continuous with outer epidermis of leaf), black with obvious reticulation (**K** 4) . . . .   1. *I. nigroreticulata*
   Velum not complete, virtually absent or reduced to a narrow rim or sometimes wider 2
2. Velum reduced to a rim about 2 mm wide (more complete in typical Angolan material); large face of megaspores with verrucae anastomosing in groups and smaller faces with low almost touching verrucae (in small megaspores often only a single verruca) . . . .   2. *I. aequinoctialis*
   Velum virtually absent or vestigial 3
3. Small rosette plant, only 2.5 cm tall with sporangia only 2.2 × 1.7 mm (**K** 4, Thika) . . . .   6. *I.* sp. A
   Larger plants, 5–60 cm tall with larger sporangia 4
4. Short plants 5–20 cm tall with leaves narrow in sporangial area (2–6 mm wide) and narrow wings . . . .   3. *I. welwitschii*
   More robust plants 15–60 cm tall with leaves broader in sporangial area (8–13 mm wide) and with broader wings 5
5. Sporangium wall pale . . . .   4. *I. schweinfurthii*
   Sporangium wall black . . . .   5. *I. alstonii*

1. **Isoetes nigroreticulata** *Verdc.* **sp. nov.** ob velum completum *I. tenuifoliae* Jermy affinis sed velo nigro reticulato differt. Type: Fort Hall District, Thika, N side of Thika R., E of Nairobi – Muranga road opposite to Horticultural Research Station turn-off, *Faden & Kabuye* 71/550 (K!, holo., EA, iso.)

Annual (fide Faden), almost completely submerged or on mud. Corm 2–3-lobed; leaves densely tufted, 3–16 cm long, 0.7–8 mm wide at middle; ligule narrowly triangular, 2 mm long. Velum complete, shiny black and reticulate, the adjacent leaf margin pale and thin, 1–2 mm wide above, narrower in sporangial area. Megaspores with rather scattered tubercles on large face, 3 smaller faces with small low indistinct tubercles or almost smooth.

KENYA. Fort Hall District: Thika, N side of Thika R., E of Nairobi – Muranga road, opposite to Horticultural Research Station, 1 July 1971, *Faden et al.* 71/544! & same locality 11 July 1971, *Faden & Kabuye* 71/550! & same locality 11 July 1971, *Faden & Kabuye* 71/571!
DISTR. **K** 4; not known elsewhere
HAB. Seasonal and long lasting pools and marshes with grasses, sedges, *Aponogeton, Eriocaulon, Crassula, Nesaea, Rotala* etc.; ± 1520 m

SYN. *I. tenuifolia* sensu Faden in U.K.W.F. ed. 2: 39 (1994), *non* Jermy

NOTE. I have examined the type of *I. tenuifolia* Jermy from Ghana (*Hall* 3728; BM, holo., K!, iso.) and it has a pale velum different in colour and sculpture. I do not think they are conspecific.

2. **Isoetes aequinoctialis** *A.Braun* in Kuhn, Fil. Afr.: 195 (1868); Jermy in F.Z., Pterid.: 30 (1970); Schelpe & Jermy in C.F.A., Pterid: 31 (1977); W. Jacobsen, Ferns S. Afr.: 155, t. 99 (1983); Schelpe & N.C. Anthony, F.S.A. Pterid.: 25 (1986); J.E. Burrows, S. Afr. Ferns: 33, illustr. 8/28, 28a (1990); Schippers in Fern. Gaz. 14: 177 (1993). Type: Angola, between Pungo Andongo and Sansamandra, near R. Cuanza, *Welwitsch* 50 (B, holo., BM, K!, LISU, iso.)

Corm 3-lobed, 4–10 mm wide, basal scales brownish, thin, papery or absent. Leaves 10–20, 4–35 cm long, 1–1.5 mm wide above, up to 3 mm just above sporangia, more or less trigonous; ligule triangular, attenuate, often 3-lobed (fide

Schelpe & Anthony). Velum covering $^1/_2$ to $^3/_4$ of sporangium or much reduced to a rim 1–2 mm wide (or less than 1 mm fide Schelpe & Anthony). Megaspores of two sizes, the large face coarsely tuberculate, the tubercles usually anastomosing; 3 smaller faces with large low very close tubercles or in small spores one single tubercle. Microspores scabrate.

TANZANIA. Tunduru District: just E of Songea District Boundary, 6 June 1956, *Milne-Redhead & Taylor* 10590!
DISTR. **T** 8; Ghana, Mali, Angola, Zambia, Zimbabwe, Namibia, South Africa
HAB. Damp sandy ground by drying water hole in boggy grassland; 870 m

NOTE. Schelpe and Jermy annotated the specimen cited as *I. aequinoctialis* but it certainly differs from the type in having a much more reduced velum; the collectors describe the sporangia as dark whereas they are pale in typical material. The lobing of the corm is not described by the collectors and not clear from the dried material. Burrows description of the sculpture of the megaspores differs totally from his figure and does not mention anastomosing.

3. **Isoetes welwitschii** *A.Braun* in Kuhn, Fil. Afr.: 196 (1868); Schelpe & Jermy, C.F.A., Pterid.: 32 (1977); Schelpe & N.C. Anthony, F.S.A., Pterid.: 28 (1986); W. Jacobsen, Ferns S. Afr.: 154 (1983); J. Burrows, S. Afr. Ferns: 34, t 4/5, illustr. 8/29, 29a, 29b (1990); Faden in U.K.W.F. ed. 2: 39 (1994). Type: Angola, Huila, Morro de Lopollo, Empalanca, *Welwitsch* 166 (B, holo., BM, K!, iso.)

Small ± terrestrial herb 5–20 cm tall with corm scarcely developed or up to 1.5 cm wide, 3(–4)-lobed. Leaves 5–35, very narrow, ± semi-circular in cross-section, abruptly enlarged to 2–6 mm wide in sporangial area; dried specimens often rather distinctly bicolored, whitish at base (sporangial part and about 1.5 cm above it) contrasting with dark green upper parts. Sporangia pale or ± dark, 5–7(–9) mm long, 3–3.5(–5) mm wide. Ligule narrowly triangular, 2.5 mm long. Velum almost absent or rarely a rim ± 1 mm wide. Megaspores of two sizes; larger face with low verrucae fairly well spaced, often large and small mixed, usually not anastomosing in Flora area but often doing so in specimens from southern Africa; 3 smaller faces with spaced tubercles confined to the centre or if more extensive then central ones the largest or sometimes ± smooth. Microspores minutely tuberculate.

KENYA. Northern Frontier District: E side of summit plateau of Lolokwi [Ol Lolokwe], 25 Mar. 1978, *Gilbert* 5002!; Trans Nzoia District: Suam Saw Mill track, edge of Kiptogot R. valley, 12 June 1971, *Faden et al.* 71/455!; Nairobi District: Nairobi, Langata South, Aug. 1977, *Gilbert* 4827!
DISTR. **K** 1, 3–5; **T** (see note); Ghana, Nigeria, Central African Republic, Sudan, Ethiopia, Eritrea, Angola, Zambia, Botswana, South Africa and Madagascar (this distribution is dependent on accepting the synonymy mentioned in the note)
HAB. Black mud of pools and forest streams, on granite rocks, seepage areas and vleis in grassland with scattered bushes; together with various sedges, *Eriocaulon*, *Ilysanthes* etc., *Aeschynomene* etc.; 1500–2400 m

SYN. *I. abyssinica* Chiov. in Atti Soc. Nat. Mat. Modena 64: 45, fig. (spores only) (1933); Faden in U.K.W.F. ed. 2: 39 (1994). Type: Ethiopia, "Amhara to Dembia", Gondar, Cococc Valley above Gondar, *Chiovenda* 1809 (FT, lecto., BM, photo.!)*

NOTE. Annotations in the BM herbarium indicate that Jermy finally considered *I. abyssinica* to be conspecific with *I. welwitschii* A.Braun and also that *I. nigritiana* Kuhn and *I. garnieri* A.Chev. & P.Monnier should be included. I have not seen any of Chiovenda's syntypes and it is possible they belong to more than one species. Pichi Sermolli refers to *I. abyssinica* in B.J.B.B. 53: 190 (1983) & 55: 197 (1985) but these Burundi specimens seem too robust as is also the case with material cited

* Chiovenda cites 9 syntypes in his original reference, all from Ethiopia and Eritrea. A photograph of *Chiovenda* 1809 at the BM shows that Jermy had annotated this as lectotype but he does not appear to have published the choice.

by Hall in Bot. J. Linn. Soc. 64: 117–139, fig. 2e, 3c, e, f, 4d, f, 5c, 6b, c (1971) from Ghana. Faden keeps up both species in U.K.W.F. ed. 2: 39 (1994) separating them as follows –

megaspores with numerous large occasionally coalescing tubercles on the convex surface; triradiate facets completely covered by uniform tubercles   *I. welwitschii*
megaspores with fewer lower tubercles that never coalesce on the convex surface; triradiate facets with tubercles confined to centre or if more extensive then the ones in the centre the largest or occasionally tubercles lacking . . . .   *I. abyssinica*

He records *I. welwitschii* from Lolokwi (15 Apr. 1979, *Gilbert* 5366) and the S end of Mua Hills (2 Feb. 1969, *Napper & Faden* 1868). Burrows' figures of the spore sculpture show very sparse verrucae quite unlike that of the type specimens.

Some specimens show characters which need investigation in the field to assess their possible importance. Several (e.g. Nairobi, 24 July 1977, *Gilbert & Gillett* 4800; Nairobi, Aug. 1977, *Gilbert* 4827; Nairobi National Park, 25 June 1971, *Lye* 6310) have dark marks on the outside of the sporangial area of the leaves (individual scattered dark epidermal cells) but initial suspicions this could be correlated with dark sporangia and velum rims ± 1 mm wide proved unfounded. *Faden et al.* 71/455 cited above has broadly triangular black corm scales with very cuspidate apices but few specimens have intact scales and the possible taxonomic significance of such characters can only be assessed in the field, but the fact that *Gilbert & Mesfin* 6500 (10 km Eldoret – Kitale, 8 Oct. 1981) is similar suggests it may be significant.

Schippers records *I. abyssinica* from **T** 3 near Muheza and from along the coast (*Wingfield* 2032). The Muheza specimen has not been seen but the Wingfield specimen is far too large, particularly the leaf length and sporangia and is here referred to *I. alstonii*.

Material from the Mathews Range also needs more study; *Gilbert* 5366 (Lolokwi [Ololokwe], 14 Apr. 1979) has sporangia 9 × 5 mm and the verrucae on large face of megaspores more oblong, slightly narrowed at the base. *Gilbert* 5002 cited above is not mature.

4. **Isoetes schweinfurthii** *Baker* in J.B. 18: 108 (1860); Alston in Mém. I.F.A.N. 50: 48, t. 8 fig. 7 (1957); W. Jacobsen, Ferns S. Afr.: 153 (1983); Schelpe & N.C. Anthony, F.S.A. Pterid.: 28 (1986); J. Burrows, S. Afr. Ferns: 36, illustr. 8/30, 30a (1990); Schippers in Fern Gaz. 14: 177 (1993). Type: Sudan, Djur, Seriba Ghattas, *Schweinfurth* 1962 (K!, holo., B, BM, FT, P)*

Corm 3-lobed or unlobed; bud scales dark brown to blackish, broadly triangular, 6 mm long, 4 mm wide, cuspidate. Leaves (5–)12–30(–40), 15–60 cm long, triangular and 1 mm wide above, 8–10 mm wide in the sporangial area and broadly winged; ligule narrowly triangular, attenuate, 2–6 mm long. Sporangia round to oblong, 5–6 mm long and wide. Leaf lacunae with small internal setae on thick wall. Velum reduced to a vestigial rim. Megaspores of two sizes; large face strongly verruculate all over, often with both small and large verrucae, the 3 smaller faces almost smooth or with a few low elevations or few small tubercles in the middle or in small spores with a single central tubercle. Microspores not seen.

TANZANIA. Tabora District: Kaliua, 18 June 1980, *Hooper et al.* 2057!; Songea District: N of Songea, by R. Luhimba, 6 May 1956, *Milne-Redhead & Taylor* 10003! and Hanga Farm, 27 June 1956, *Milne-Redhead & Taylor* 10918!
DISTR. **T** 4, 8; Morocco, Senegal, Ivory Coast, Central African Republic, Sudan, Angola, Zambia, Mozambique, Zimbabwe, Namibia, South Africa and Madagascar (distribution from Burrows)
HAB. Seasonally flooded pans (growing in 13 cm of water or sometimes ± submerged) or boggy rice cultivations; 1000–1050 m

* Schelpe followed by Burrows has chosen the B sheet as lectotype, but the species was described by Baker based on a ms name of A. Braun who died in 1877. The K sheet is the holotype.

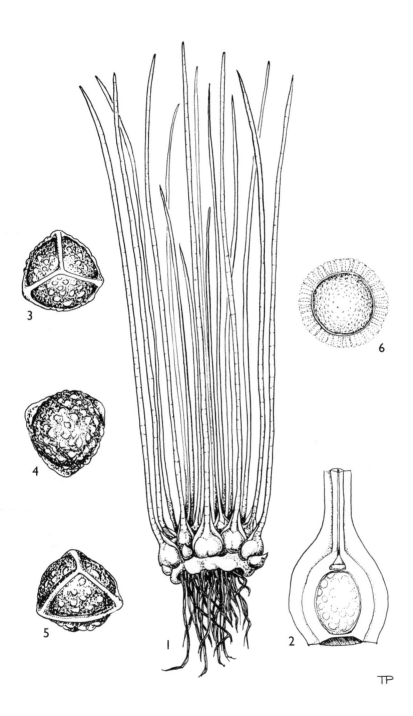

FIG. 1. *ISOETES ALSTONII* — **1**, habit, × ²/₃; **2**, section of sporangium/leaf base, × 48; **3**, proximal view of megaspore, × 48; **4**, distal view of megaspore, × 48; **5**, lateral view of megaspore, × 48; **6**, proximal view of miscospore, × 900. All from *Leach & Rutherford-Smith* 10952. Drawn by TP, and reproduced with permission from Flora Zambesiaca.

Syn. *I. rhodesiana* Alston in Bol. Soc. Brot. sér. 2, 30: 17 (1956); Jermy in F.Z. Pterid.: 30 (1970); W. Jacobsen, Ferns S. Afr.: 353, fig. 79 (1983). Type: Zimbabwe, Nyamandhlovu district, Bongola Pasture Substation, in water storage tank, the clay bottom of which had been obtained from a nearby pan in mopane woodland, *West* SRGH 30263* (BM, holo., K!, SRGH, iso.)

    I. alstonii sensu Schippers in Fern Gaz. 14: 177 (1993) pro parte, *non* Reed & Verdc.

Note. Alston does not describe the corm of *I. rhodesiana* but Jacobsen describes it as 2-lobed from further collections in Zimbabwe. Milne-Redhead & Taylor describe the corm of their 10003 as unlobed but do not describe that of their 10918; that of their 9212 (Songea District, 9.5 km SW of Songea, Kwamponjore Valley) is described as usually 3-lobed. The character is of dubious value.

5. **Isoetes alstonii** *Reed & Verdc.* in Kirkia 5: 19, t. (1965); Launert in Prodr. Fl. S.W. Afr. 1: (unpaginated) (1969); Jermy in F.Z. Pterid.: 30, fig. 6 (1970); W. Jacobsen, Ferns S. Afr.: 154, figs. 98a, 98b (1983); J. Burrows, S. Afr. Ferns: 36, t. 5, fig. 1, illustr. 8, fig. 31, 31a (1990); Schippers in Fern Gaz. 14: 177 (1993). Type: Zimbabwe, Victoria Falls, S bank in front of main falls, *Greenway & Brenan* 8012 (EA!, holo., BM, K!, PRE, SRGH, iso.)

Corm 1–2 cm in diameter, 2-lobed (fide Burrows). Leaves 30–40, caespitose, 15–40 cm long, about 1–2 mm wide above but basal part in sporangial area ovate-oblong, 10–16 mm long, 8–13 mm wide with hyaline margins; ligule triangular, 2–3 mm long, 1.5 mm wide; velum reduced to narrow rim. Leaf lacunae with small setae present in some specimens. Sporangium oblong to ellipsoid, 5–16 mm long, 3–6 mm wide, drying black and shiny. Megaspores of two sizes, usually triradiate, rarely quadriradiate, large face densely verrucose, sometimes small and large verrucae mixed, slightly anastomosing, 3 smaller faces with smaller verrucae in centre or granular with low indistinct verrucae or with a single large verruca. Microspores winged, reticulate (fide Jermy). Fig. 1 (page 5).

Tanzania. Ulanga District: Mlahi, 17 May 1977, *Vollesen* MRC 4586!; Songea District: waterfall on R. Luhira, 18 Mar. 1956, *Milne-Redhead & Taylor* 9244!; Masasi District: NE of Masasi, Pangani Hill, 11 Mar. 1991, *Bidgood et al.* 1919!

Distr. **T** 6, 8; Zambia, Mozambique, Zimbabwe, Namibia and Madagascar

Hab. Damp pockets of soil along streams and in rocky river beds, submerged only at very wettest part of season, also in small temporary waterholes in wooded grassland with 20–40 cm of water; 0–1050 m

Syn. *I. rhodesiana* sensu Vollesen in Opera Bot. 59: 108 (1980), *non* Alston

Note. The status of this plant is very uncertain. Alston was convinced it was a distinct species. Jermy (in F.Z.) suggested that *I. rhodesiana* might be a habitat form of *I. alstonii*. Schelpe and Anthony sink both into *I. schweinfurthii* but annotations in the BM herbarium (e.g. of *Gilbert & Thulin* 700 from Ethiopia) show that in 1979 Jermy and Schelpe felt *I. alstonii* should be considered a subspecies of *I. schweinfurthii*; this was never published. Burrows is convinced that *I. alstonii* is distinct because of its 2-lobed corm, blackish sporangia and denser verrucae on megaspores. No mention was made of corm lobing in the original description of *I. alstonii*, or in the field notes of the material I have seen from the Flora areas. *Hooper & Townsend* 2109 (Tabora District, 8 km S of Kaliuwa, 22 June 1980) of which only spirit material has been seen would appear to belong here, but might have been collected to provide spirit material of what was thought to be the same species as *Hooper & Townsend* 2057 cited above under *I. schweinfurthii*.

* F.Z. gives *West* 3075, presumably the collector's own number.

6. **Isoetes** sp. A

Small rosette plant with leaves only 2.5 cm long, some very narrow and ± 0.2 mm wide, others up to 0.8 mm wide, abruptly widening at base to ± 4 mm wide with hyaline margin 1 mm wide, sometimes with a blackish band above sporangium and strongly reticulate cells in this area on both faces; ligule narrowly triangular, 1 mm long. Velum reduced to a very narrow rim. Sporangium rounded-oblong, 2.2 mm long, 1.7 mm wide. Megaspores white with dense highly raised ± oblong tubercles on all faces, somewhat smaller on the smaller faces.

KENYA. Fort Hall District: Thika, N side of Thika R., E of the Nairobi – Fort Hall road opposite turn off to the Horticultural Research Station, 11 July 1971, *Faden & Kabuye* 71/551!
DISTR. **K** 4
HAB. Seasonal pool with many sedges and *Eriocaulon*; ± 1520 m

NOTE. It does not seem possible this is a depauperate form of *I. welwitschii* growing in rather drier conditions. Further study of more material may indicate it is a distinct species.

UNCERTAIN SPECIES

Isoetes sp. B

Mention of this species must be made since Faden (in litt.) states it is "obviously different from any other Kenyan species (larger size, internal cilia in leaf lacunae)".

KENYA. Tsavo District: Tsavo area, Mayer's Ranch, *Faden* et al. 72/148
DISTR. **K** 7
HAB. 500 m

NOTE. A specimen of this was sent to Jermy but the collection is not at K or BM, possibly in both cases loaned to US. It is possibly *I. schweinfurthii* or *I. alstonii*.

INDEX TO ISOETACEAE

New names validated in this part

Isoetes nigroreticulata *Verdc.* **sp. nov.**

PLANTS PEOPLE
POSSIBILITIES

First published in 2005 by
Royal Botanic Gardens, Kew
Richmond, Surrey, TW9 3AB, UK
www.kew.org

ISBN 1 84246 112 5

Design by Media Resources, typesetting and page layout by Margaret Newman, Information Services Department, Royal Botanic Gardens, Kew.

Printed by Cromwell Press Ltd.

For information or to purchase all Kew titles please visit www.kewbooks.com or email publishing@kew.org